REALLY STRANGE ADAPTATIONS

REALLY STRANGE MARINE ANIMALS

CAITIE MCANENEY

PowerKiDS press
New York

Published in 2017 by The Rosen Publishing Group, Inc.
29 East 21st Street, New York, NY 10010

First Edition

Editor: Theresa Morlock
Book Design: Reann Nye

Photo Credits: Cover, p. 1 (shark) nicolasvoisin44/Shutterstock.com; cover, pp. 1–32 (background) Andrey_Kuzmin/Shutterstock.com; p. 4 (coral reef) aquapix/Shutterstock.com; p. 4 (deep sea) Ethan Daniels/Shutterstock.com; p. 5 Franco Banfi/WaterFrame/Getty Images; p. 7 © KerrynParkinson/NORFANZ/Caters News via ZUMA Press; p. 8 Doug Perrine/Photolibrary/Getty Images; pp. 9, 30 (anglerfish) superjoseph/Shutterstock.com; p. 10 Derek Heasley/Shutterstock.com; pp. 11, 30 (great white shark) Mogens Trolle/Shutterstock.com; p. 11 (tiger shark) Michael Bogner/Shutterstock.com; p. 11 (whale shark) Amanda Nicholls/Shutterstock.com; p. 12 (top) https://commons.wikimedia.org/wiki/File:Mistukurina_owstoni_museum_victoria_-_head_detail.jpg; p. 12 (bottom) Dorling Kindersley/Getty Images; p. 13 (megamouth shark) TORU YAMANAKA/AFP/Getty Images; p. 13 (background) Creative Mood/Shutterstock.com; p. 14 https://commons.wikimedia.org/wiki/File:Giant_Oarfish.jpg; pp. 15, 30 (oarfish) Jonathan Bird/Photolibrary/Getty Images; p. 16 (coconut octopus) Andrea Izzotti/Shutterstock.com; p. 16 (common octopus) Vladimir Wrangel/Shutterstock.com; p. 16 (Caribbean reef octopus) Ethan Daniels/Shutterstock.com; pp. 17, 30 (blue-ringed octopus) YUSRAN ABDUL RAHMAN/Shutterstock.com; p. 18 (normal pufferfish) Isabelle Kuehn/Shutterstock.com; pp. 18 (scared pufferfish), 30 (pufferfish) Beth Swanson/Shutterstock.com; pp. 19, 30 (stonefish) Yann hubert/Shutterstock.com; p. 20 https://commons.wikimedia.org/wiki/File:Scotoplanes_globosa_and_crab.jpg; p. 21 metha1819/Shutterstock.com; p. 23 Dave Fleetham/ Perspectives/Getty Images; p. 24 Luis Díaz Devesa/Moment Open/Getty Images; p. 25 Marevision/age fotostock/Getty Images; pp. 26, 30 (pipefish) Hendrik Martens/Shutterstock.com; pp. 27, 30 (leafy dragon) Tony Wear/Shutterstock.com; p. 29 timsimages/Shutterstock.com.

Cataloging-in-Publication Data

Names: McAneney, Caitie.
Title: Really strange marine animals / Caitie McAneney.
Description: New York : PowerKids Press, 2017. | Series: Really strange adaptations | Includes index.
Identifiers: ISBN 9781499427387 (pbk.) | ISBN 9781499428537 (library bound) | ISBN 9781508153016 (6 pack)
Subjects: LCSH: Marine animals–Juvenile literature. | Marine animals–Miscellanea–Juvenile literature.
Classification: LCC QL122.2 M367 2017 | DDC 591.77–dc23

Manufactured in the United States of America

CPSIA Compliance Information: Batch #BW17PK: For Further Information contact Rosen Publishing, New York, New York at 1-800-237-9932

CONTENTS

AMAZING ADAPTATIONS

Most of Earth's surface is covered in water. In fact, oceans cover more than 70 percent of our planet. Oceans are full of many different **habitats**, from coral reefs to the deep sea. Many of these habitats are **extreme**. Some are dark and cold, while others are well lit and tropical. Because of these differences, there is a great **diversity** of animal species in the oceans. Some scientists believe there are millions of species there, and we don't even know half of those species.

Many ocean creatures, such as this wobbegong shark, almost look unreal. The wobbegong uses **camouflage** to blend in with its surroundings.

Each **marine** species has adapted to its habitat over thousands, or even millions, of years. Some rely on color, shape, or other special **defenses** and skills to help them survive. Read on to learn about some strange marine animals!

IT'S THE BLOB!

Many argue that the blobfish is the world's ugliest animal. In fact, it was given that title in 2013 from the Ugly Animal Preservation Society.

The blobfish is part of a family group called fatheads. It has little muscle or bone. The blobfish lives in deep ocean waters where pressure is very high. The jellylike flesh of the blobfish is an adaptation to help it deal with this pressure, allowing it to float above the ocean floor.

It doesn't have the muscle strength to chase and catch prey. Scientists believe it just opens its mouth and lets small ocean creatures float in. This may be the laziest fish in the ocean!

Blobfish grow to be about 1 foot (0.3 m) long. They live off the coast of Australia and prefer deep waters. There's no need to look pretty when you're swimming in darkness.

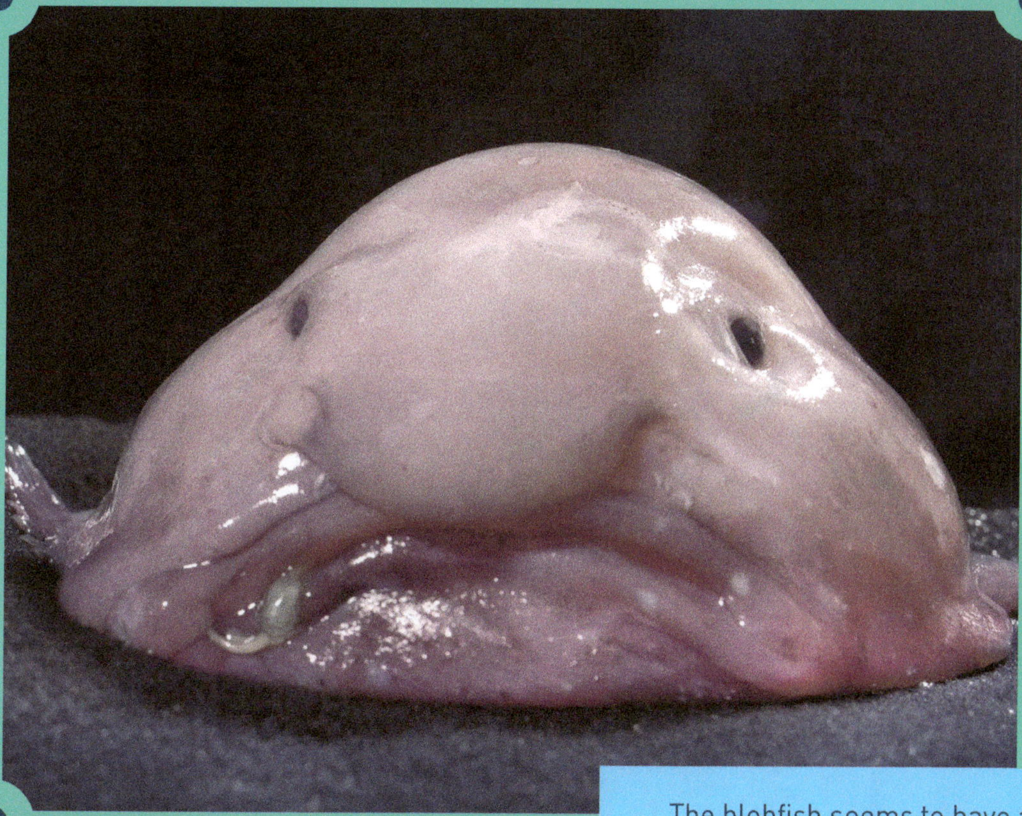

The blobfish seems to have a constant frown on its face.

ANGLERFISH

The deeper you go in the oceans, the more **bizarre** the animals seem. Humans don't usually see animals from the deep sea, but sometimes they wash up on shore. This gives us a glimpse into life in the ocean's depths.

Some deep-sea creatures adapt to the darkness by letting off light, which is an adaptation called bioluminescence. The anglerfish uses its adaptation to attract prey and mates.

The anglerfish may be one of the strangest deep-sea animals. There are more than 200 species of anglerfish, and most live in the Atlantic and Antarctic Oceans. Females are much larger than the males, and they can grow to more than 3 feet (0.9 m) long. The females have an interesting adaptation called the **photophore**, which is an organ that gives off light. It hangs at the end of a body part that looks like a fishing pole. The photophore lures in smaller fish. The curious small fish see the light, swim in for a closer look, and are gobbled up by the anglerfish.

SUPER SHARK SENSES

Sharks are some of the greatest predators in the oceans. Some scientists believe sharks have roamed the oceans for about 450 million years. There are more than 400 species of sharks, and each has amazing adaptations that keep them alive and on top of their food chain.

Most sharks have the same basic senses. Many can see well, even in low light, and they also have a great sense of smell. Scientists believe a great white shark can smell one drop of blood in 10 billion drops of water.

HAMMERHEAD SHARKS

Hammerhead sharks are huge predators that can grow up to 20 feet (6 m) long. They also have some of the strangest-shaped heads in the animal world. This is actually an adaptation. Hammerheads can use their heads to pin stingrays against the ocean floor. Their eyes are set far apart from one another, which gives them a great range of sight. They also have a better sense of electrical fields than other sharks because of their head shape.

great white shark

Some sharks use their supersenses to hunt large prey. Others feed on tiny ocean animals and plants.

tiger shark

whale shark

One shark supersense is the ability to sense electrical fields. That helps them feel even the smallest movements of living things around them. Sharks use this sense to guide them when they migrate across the ocean. Prey animals don't stand a chance!

DEEP-SEA SHARKS

The goblin shark is a rare shark that lives in the deep sea. It has a long, pointed snout and a mouth full of thin, sharp teeth. Goblin sharks inch slowly toward prey. Then, the goblin shark's jaw darts out of its mouth and grabs the meal. The shark's ability to "disconnect" its jaw comes from flaps of skin on the snout.

goblin shark

Sharks in the deep sea can look like something out of a nightmare!

Another deep-sea shark with an unusual mouth is the megamouth shark. This shark is rarely seen. It can grow to more than 20 feet (6 m) long. This shark's mouth has 50 rows of teeth. Luckily for us, the megamouth uses its big mouth to catch as many tiny ocean animals as it can. Like the goblin shark, the megamouth can push out its jaw.

GIANT OARFISH

The ocean is a big place, and it's home to the world's biggest animals. For hundreds of years, there have been legends of huge sea monsters resembling snakes. These sea serpent legends may be based on the world's longest bony fish—the oarfish. Thankfully, this sea monster would rather eat tiny **plankton** than people.

Oarfish are rarely seen in the wild because they live so far underwater. They live in depths between 650 feet and 3,200 feet (198 and 975 m). When oarfish swim, their long, thin bodies look like silvery ribbons. Oarfish often float vertically, probably to camouflage themselves in the water. These fish can grow to amazing lengths. The longest reported oarfish was 56 feet (17 m) long.

Sometimes oarfish wash up on land and give people a closer look at their huge, silvery bodies.

BLUE-RINGED OCTOPUS

The blue-ringed octopus may be small, but don't let its size fool you. It's extremely **venomous**. Of all the animals in the octopus or squid family, this is the most deadly. The bite of a blue-ringed octopus could kill 26 adults.

OCTOPUS ADAPTATIONS

Octopuses are the smartest **invertebrates** in the world. They can complete mazes, escape from **enclosures**, and unscrew jars. Octopuses are even known to play. Octopuses are built for survival. They have a large brain for their size, three hearts, and sharp eyes. They have eight arms, and each one is lined with suckers to grip prey and feel around. Some octopuses have ink to spray at predators, which scares the predator and pushes the octopus away from it.

A blue-ringed octopus is only about the size of a golf ball. It doesn't want to hurt people, but people can get hurt if they pick up the octopus or step on it.

The blue-ringed octopus's venom is an amazing adaptation. However, they have other adaptations, too. Their eight **tentacles** help them grab things. Under their body is a hard beak they use to bite and chew. One of their coolest adaptations is their glow. When the blue-ringed octopus is resting, its color is tan or yellow so it can blend in with the sand. However, if the octopus is bothered, rings on its skin glow bright blue. That's its way of telling the predator to stay far away.

STINGING STONEFISH

The stonefish is one of the most venomous fish in the world. But this fish isn't just venomous—it's also a master of camouflage. These two adaptations make the stonefish dangerous to both prey and people who accidentally step on it.

scared pufferfish

pufferfish

THE POISONOUS PUFFERFISH

Pufferfish are also at the top of the world's most poisonous list. When they're scared, they inflate like a funny-looking, spiky balloon. Pufferfish may not look scary, but this fish is very poisonous when eaten. There's enough poison in one pufferfish to kill 30 people. Some pufferfish are only 1 inch (2.5 cm) long, while others can grow to more than 2 feet (0.6 m) long. In Japan, pufferfish are made into an expensive dish called fugu. Fugu needs to be made correctly, or the person eating it could die!

You might think a stonefish is just a bunch of rock or coral, until you see this grumpy face in the sand.

Stonefish are found in shallow waters in northern Australia. They lie on the ocean floor and cover themselves in sand. Their skin looks rocky or wart-covered, and it's brown or gray. Most animals or people wouldn't be able to tell this fish from the coral or rocks around it.

The stonefish delivers its deadly venom through spiky spines on its back. When the fish feels threatened, it lifts its spines, which sting anything that comes near. The venom can kill both animals and people. A person can take **antivenin** to treat stonefish venom.

SEA CUCUMBERS

Sea cucumbers get their name from their odd shape. Most look like a spiky cucumber. These wormlike marine animals make their home on or near the ocean floor.

Sea cucumbers are like the earthworms of the sea. They act as **decomposers** for their ecosystem. They feed on tiny ocean animals and plants, and even waste that floats by. Their mouths are surrounded by tentacle-like feet that gather up food. Each sea cucumber has up to 30 of these tube-shaped feet. The sea cucumber breaks down its food and releases smaller pieces back into the ocean.

SEA PIG

One species of sea cucumber is called the sea pig. The sea pig is a pink, round sea cucumber. Like other sea cucumbers, they're full of liquid, which gives them their balloon-like shape. Sea pigs live on the ocean floor in the deep sea. They "walk" on the ocean floor with their tube feet, which are longer than those of most sea cucumbers. Longer tubes on their back help them feel around the floor. Tubes near the "mouth" search out food particles that are hidden in the mud.

Sea cucumbers don't look much like living animals at all. However, they're very important to their ecosystem.

Sea cucumbers have one of the strangest defenses in the ocean. When threatened, the sea cucumber can push out its internal organs. This might scare the attacker or convince it to leave the cucumber alone. Then, the sea cucumber grows new organs.

UNICORN OF THE OCEAN

Do unicorns exist? Maybe horned horses don't exist, but horned porpoises do. Male narwhals have a long, pointed tooth that looks almost like a sword. This tusk grows through the narwhal's upper lip. Some narwhals have tusks that can grow up to nearly 10 feet (3 m) long. The tusk has millions of nerve endings inside. Some scientists think it may help narwhals learn about their surroundings.

Narwhals can grow up to 20 feet (6 m) long. They live in the cold waters of the Arctic, where they eat fish and shrimp. They live deep under the icy surface, coming up only for air.

One of the narwhal's greatest adaptations is its connection to other narwhals. Sometimes there are dozens of narwhals swimming together, and sometimes there are hundreds. This helps them hunt and stay safe from predators.

Scientists don't know why male narwhals have a tusk. Some believe they use it to attract females.

UNDERWATER VAMPIRE

Sea lampreys are the real vampires of the ocean. These long, thin fish look a lot like eels. They can grow up to 20 inches (50.8 cm) long.

Sea lampreys can have more than a hundred teeth in their mouth.

Sea lampreys may look harmless—until they open their mouths. They have circular mouths filled with tiny teeth. The teeth sink into fish flesh and hold on, sucking blood. Because lampreys live off other animals, they're called parasites. They kill many fish and wound many others. They don't just take a quick bite; they might keep sucking a fish's blood for weeks!

Sea lampreys used to live only in the ocean. However, they invaded the waterways of the Great Lakes region and killed off many native fish. You can tell when a fish has been attacked by a lamprey because of the round wound a lamprey's mouth leaves.

MASTER OF DISGUISE

Leafy sea dragons are masters of camouflage. Not only do they match the color of their surroundings, they're also shaped like objects from their habitat.

Found along the southern coast of Australia, leafy sea dragons often have thin brown or yellow bodies. They have leaf-shaped body parts that help them look just

PIPEFISH

Like leafy sea dragons, pipefish are great at camouflage. These long, thin fish have tubelike bodies. Like sea dragons and sea horses, pipefish have bony plates covering their body. They can live almost anywhere, from salt water to freshwater and from coral reefs to the deep sea. Pipefish blend in with sea grasses because of their shape and size. Their color depends on the species. Some are colorful and have patterned skin, while others blend in with dull surroundings.

A predator on the hunt might not see a leafy sea dragon or a pipefish.

like seaweed. Leafy sea dragons choose their homes accordingly—they love living in sea grass meadows and seaweed beds. Leafy sea dragons also tend to drift about in the water, which makes them look like objects instead of animals. This keeps predators away.

Weedy sea dragons are very similar, but they have fewer "leaves" and are darker in color. Sea dragons are related to sea horses and pipefish.

WEIRD AND WONDERFUL

The underwater world is home to millions of amazing animals. With such a variety of habitats, it's no wonder that there's a wide variety of life. From the blobfish with its jellylike body to the narwhal with its tusk, marine life is both weird and wonderful!

Marine animals are built to survive in their underwater habitat. Some animals that live in the deep sea give off their own light. Some animals blend in with their surroundings so predators won't find them. Others have deadly venom that can kill any animal that comes too close. Less than 5 percent of the ocean has been explored, and some scientists think two-thirds of marine species have yet to be found. How many more amazing species do you think exist in the deep blue sea?

Coral reefs are colorful cities full of weird and wonderful marine animals.

UNDERWATER ADAPTATIONS

SIZE	sharks, oarfish
CAMOUFLAGE	bule-ringed octopus, stonefish, leafy sea drdgon, pipefish
VENOM/POISON	bule-ringed octopus, pufferfish, stonefish
SUPERSENSES	sharks
GLOW	anglerfish, bule-ringed octopus

Check out the amazing underwater adaptations of the animals in the book!

GLOSSARY

antivenin: A drug that stops the effects of an animal's venom.

bizarre: Very unusual and strange.

camouflage: The color or shape of an animal that allows it to blend in with its surroundings.

decomposer: An organism that breaks down dead things or waste and returns the nutrients to the earth.

defense: A feature of a living thing that helps keep it safe.

diversity: The quality or state of having many different types.

enclosure: An area that's surrounded by walls to hold an animal in.

extreme: Great or severe.

habitat: The natural place where an animal or plant lives.

invertebrate: An animal that lacks a spine.

marine: Found in the sea.

photophore: An organ that lets off light.

plankton: Small living organisms that drift in water, usually near the surface.

tentacle: A long, thin body part that sticks out from an animal's head or mouth.

venomous: Able to produce a liquid poison that is forced into prey through fangs or stingers.

INDEX

WEBSITES

Due to the changing nature of Internet links, PowerKids Press has developed an online list of websites related to the subject of this book. This site is updated regularly. Please use this link to access the list: www.powerkidslinks.com/rsa/rsma